This Fashion Fairytale belongs to

Princess _____,

whose _____ years of fearless fashion

expressions will be rewarded

with love, laughter

and lots of living.

DIANE von FURSTENBERG

and the tale of

THE EMPRESS's NEW CLOTHES

a fashion fairy tale memoir

DIANE von FURSTENBERG

and the tale of

THE EMPRESS'S NEW CLOTHES

Camilla Morton

ILLUSTRATED BY DVF STUDIO

itbooks

AN IMPRINT OF HARPERCOLLINS PUBLISHERS

To DIANE,

Who inspired this story & continues

to empower women the world over,

with thanks and much love,

Camilla x

Thank you Camilla

love is life

Diane ♡

nce Upon a Time there was a little girl with curls as dark as night who dreamt of adventure. Some girls dream of becoming a Princess—crowned with long straight golden hair, like those in classic fairy tales. For this little girl becoming a Princess was only just the start of her story . . .

While sometimes a fairy tale wish comes true, sometimes a life is even blessed with a few.

Our story begins in Brussels, where this little girl was born. It was as if she had come to restore sunshine and joy to those around her, most particularly her mother. Here was a girl full of love, laughter, hope and promise . . . and her name was

The long days seemed to drag endlessly in Diane's sleepy little town, so she passed the time inventing a world of make-believe featuring princesses, comtesses and, of course, valiant knights in shining armor.

Young Diane never wanted to stay at home and play with dolls.

She much preferred to live the fantasies and explore all the world had to offer.

On days when she was indoors she loved discovering new worlds through books. She would beg her mother to read poetry to her, then memorize and recite the verses to others. And, when there were no poems at hand, she could be found talking to herself in the mirror, pretending she was performing on stage.

This was a child who was *very* impatient to grow up, and to see and experience all the wonders life had to offer firsthand.

Over in a distant land, quite some time later—in fact far more fashion seasons into the future than you can count on one hand—a young Empress took to the throne. This was a royal who simply *loved* clothes. In fact, her existence revolved around the whole haute couture of new designers, new handbags, and looking fabulous—after all it was her "job" as ruler to be noticed and noticeable, and she already had an entire office dedicated to all her press cuttings and covers.

Every morning it was the same; she woke up, glanced at her agenda, and as she entered her enormous designer-label-stuffed closet, the first question on her mind was always the same:

Whatever shall I wear?

When she logged onto her computer, the second and third royal questions were equally predictable: *What's new?* and *What will I buy today?* And alas, these laments were often followed by a stamp of the royal foot, and the cry "I want it *now!*" It was a full-time job being a dedicated royal follower of fashion.

Fashion, it was fair to say, was her passion.

Morning, noon, and night. She had inherited this—if nothing else—from her much beloved and mourned mother. The ingenious young Empress was, in fact, so obsessed she would cram as many "official" meetings and events into a single day, not through any sense of duty but more so she could have ample excuse to change outfits as often as possible.

The whole Kingdom buzzed about their monarch's insatiable appetite for the finer things in life and everyone— from the Minister of Transport ("High Heels or Limousine shoes today, ma'am?") to the Minister of Defense

("They tried to take your place on the Alligator Bag wait list, but we arrested those ahead of you")—certainly knew how best to phrase their reports to get her attention . . .

Indeed everything went along swimmingly as long as the Empress, and her Kingdom, felt she was at the cutting edge of style.

But back to Diane's life, which was now brimming with the kind of adventure she had always craved. She'd been sent off to boarding school in Switzerland, where she found a best friend who had the long, straight golden hair that *almost all* fairy tale heroines possess. Together the two girls happily whiled away the hours laughing, telling stories, confiding in each other their hopes and dreams, and playing their favorite game of all: "Princesses."

One day, quite abruptly, Diane was transferred to yet another school in the English countryside. It was hoped that Diane's peaceful new surroundings would let her grow in idyllic calm and, more than anything, inspire a passion for nature and the arts.

Diane dreamt and wrote about love,

surely the most worthy and important adventure of them all? The lavender, leaves, rivers and ancient treasures all seemed to be tied together with ribbons of romance and the verse of great poets. She penned hundreds upon hundreds of letters from the heart and sent them out on the breeze, hoping her destiny would find her. Diane's romantic soul would always seek love, give love, and live to be in love and here she started to find her voice.

The quiet rural life in England led next to the balmy, vibrant city of Madrid, where Diane had secured a place at the University.

During her first Christmas break there she traveled to Gstaad. Here, at last, she found all the glamour and excitement she had spent her whole life seeking.

The Jet Set.

When the new school term started, she was determined to *experience* life, not just learn about it, and she knew just where to find it. . . . Her eyes had been opened to where, and how, she would thrive. So quick-thinking Diane pretended to be suffering from appendicitis, fled to Geneva, perfected her tan, and it was there she would meet that Prince.

If only life were as "grand" for our Empress, who sighed a sad little sigh as she gazed at herself in the mirror. She gazed past the crown, jewels and finery that could not conceal her sadness. She was sure, under the harsh light and delicate makeup she wore, that she could already spy lines forming around her eyes and furrowed brow. Honestly, it was this job, she'd be going gray next—surely she was far too young to need to start scheduling royal Botox?

She hadn't always been this vain, but now every time she gazed into the Imperial looking glass she saw traces of her mother's face . . . and knew she'd never equal her legacy.

When her mother died the palace had sunk under a mournful veil of black where even the sunshine refused to shine. Everything, and everyone, changed; and in desperation to find herself the Empress's obsession with fashion began in earnest. Couture had once been her mother's passion, not hers, but somehow now it made her feel "connected" to her mum—and if that's what it took to cheer up the Kingdom, and ease the constant dull ache in her heart, then shop she would.

But alas, retail therapy only went so far. Her father, the Emperor, wouldn't speak or even see his daughter. Overwhelmed with his own grief, he had relinquished his crown and all Imperial duties to her, and now, when she needed him the most, the Emperor rarely left the confines of his sorrow-drenched chambers.

The death of her mother, the Empress, had cast a long shadow over the entire Kingdom. No one seemed to notice that the lonely little girl who had just lost her mummy was now juggling the all-important role of Imperial Empress. She had no family to turn to and a Kingdom relying on her.

This was why she turned to fashion. She wanted to be noticed, to be respected, and to keep the memory of her mother alive, too. The former Empress had been such a renowned beauty and style icon, it seemed a befitting solution to the young Empress's problem. So that is what she did. . . . She would look the part. So the heir to the throne started to shop.

AND
SHOP.
AND
SHOP.

Once in a blue moon her father would see her, nod appreciatively at a dress she was wearing, and faintly smile. Then he'd vanish, back under his dark mantle of heartbreak, leaving her to worry about the land and its people. *But*, for that brief moment, his pleasure made all the frills, fashion and uncomfortably corseting seem worthwhile.

Clothes were the young Empress's only friend.

She was growing into a beautiful young woman. But her courtiers hadn't noticed. They tutted reproachfully and still thought of her as a frivolous girl playing dress-up.

Alas, she who wore the crown rarely had time for fun like others her age. How she longed for someone to take her *seriously*. She knew the time had come to do something drastic, and the Empress wanted to make sure she did something that got everyone's attention.

"Fun" was one word that happily was *definitely key* in Diane's job description.

From Geneva she had landed in Paris, where she was now working as a photographer's assistant. All around there was glamour and beauty. The flamboyant studio was filled with the hottest models, designers, artists, actors and dancers of the day. Creatives and inspiration were everywhere, it was empowering and exciting—opportunities were all around.

It was perfect . . . except for one thing. That Prince.

Her Prince.

She had met him as snow fell and another year ended. They had celebrated her twentieth birthday together, but since then they hadn't seen each other once. It was as if he had simply vanished.

A real-life Prince.

The thought of it seemed too much like a fairy tale. Diane tried to put him out of her mind, until one evening there he was at a dinner party. And she just knew . . .

Diane quickly got an advance of one month's salary, tucked the studio radio into her bag, and, before she had time to regret her decision, was on a flight following her heart to St. Moritz. The feeling was mutual and this time her Prince was determined not to let the exotic beauty escape him. When Diane's next birthday rolled around his gift was a plane ticket to New York City, where she joined him three weeks later . . . and she stayed for two months.

Diane would have lingered longer, but she didn't want her Prince to think that she was anything but independent. She had so many ambitions, so many ideas, and so many things she wanted to see and do . . . so she forced herself out of her

love bubble. Diane knew what she wanted to do and now was time to act on it. She was going to do something to empower women, and this took her to Italy, where she became a fashion apprentice—learning about spinning and knitting, the printing process, and pigments of all kinds.

But it wasn't long before the apprentice became the teacher. With her bold style, modern attitude, ideas and daring sense of color, she was unstoppable—no wonder her Prince came to find her.

Unlike Diane, the young Empress was very skeptical about romance, but for fashion she was anyone's fool. She would try the wildest of styles just to make her father smile. Every crazy fad, every new designer name and accessory seemed irresistible to her; the more outlandish the outfit, the more liberated she felt in it. Fashion empowered her far more than even the heaviest of her crowns. She was a leader in every sense, and the higher her heels, the more she felt people looking up to her. If only her parents could see her now . . .

Then, one day two curious-looking gentlemen were presented to her at court.

Normally such eccentrics wouldn't have got an audience, but they said they had journeyed all the way from Europe, which she knew to be the home of haute couture. . . . It was on her list of places to visit but she always had so much to attend to here in the Kingdom that she hadn't been able to go yet.

Perhaps this was why she hadn't heard of her special guests . . . ? This duo boasted that they were the hottest, most feted newcomers on the block. In fact, they were so hot that their names hadn't even hit the blogosphere yet. They didn't do shows, they didn't do the red carpet—they only dressed those in the know and those *they* deemed iconic enough.

Wow . . . that caught the Empress's interest. Instantly, she was hooked. She would overlook the cruel streak that flashed like a knife's blade across the face of the more portly of the pair as he looked her up and down. She would also ignore how her stomach flipped with misgiving when the man with the long crooked nose and the spindly spider-like fingers spoke.

Neither of these men looked like "designers" to her. Perhaps she was losing her edge? They were obviously very avant-garde. . . . She had clearly spent too much time worrying about the Kingdom. Maybe *they* could create something for her to wear to the upcoming ceremony in honor of her mother? With that hope she decided to wave aside any bad feelings and listen to their tale.

She just *had* to find a way to get on their client list . . .

so she gestured for the pair to talk . . . and talk . . .

The Empress was close to nodding off as the portly one droned on. She distractedly gazed at their lurid matching striped Breton tops—so that was "Euro Chic," she mused—while they listed their many accomplishments. . . . She was only saved from the endless soliloquy by a din of golden goblets crashing to the floor from their tray as

26

a footman, who looked similarly befuddled, hurriedly tidied up his clumsiness.

The crooked-nosed one ignored the commotion. He knew he needed to regain the fleeting royal attention fast, so he knelt beside the Empress and began speaking in hushed tones as if casting a spell. The stale stench of his breath overwhelmed her. She wanted to hide under a handkerchief but instead listened to the secret elixir he whispered ever so softly in her ear. "Majesty," he said. "Sometimes, for the very, very VIP, we can make a special cloth, not like these common threads you've been subjected to wearing . . ."

The Empress coughed and her guards flinched. She wasn't going to like that . . . but instead of exploding she suppressed her instinct to speak. No one ever dared to challenge her anymore. Plus, she was curious—she had always been rather proud of her fashion sensibility, and *certainly* didn't like to think of her priceless couture pieces as "common." The Empress decided she would bite her tongue, bat her eyes, take a deep breath and flash her most hypnotic and persuasive smile. She simply had to gain the favor of these strange fashion gurus.

"Go on," she purred. "Few really know how I suffer. I am so over mink. Silks and chinchilla are passé, don't you think? It is high time designers got with the groove. Creatively I've been so bored, but I've a feeling you might be just the men to help me. Tell me what news from Paris?!"

The pair smiled oily smiles that would have stopped a songbird mid-flight. . . . They had her. They then told the impressionable young Empress that they were inclined, after meeting her, to take her on as their client.

The Empress nearly fainted with gratitude. She had no idea how "out" her exquisite gowns would be considered in Paris, or how perilously close she had come to committing a regal fashion faux pas. Thank goodness this strange duo had arrived just in the nick of time. Her father would never have forgiven her if she had worn something less than spectacular on the day all the Kingdom gathered to remember her mother.

Diane was also celebrating: she and her Prince were newly married and were now expecting their first child. In truest fairy tale style their love seemed to defy all the odds; the young Jewish girl from Belgium had fallen in love and married a Noble German Prince.

Diane returned to the factory glowing with happiness and announced, "I'm pregnant, I'm married and I'm off to America—let's get a collection together!"

Inspired by her energy and enthusiasm, her team did just that.

Several months later, the joyous couple sailed into New York harbor, past the Statue of Liberty, with all of Diane's new dress samples folded neatly in her trunk.

But the Princess Bride did not have royal duties on her mind— she was here for business as well as pleasure. Diane began tirelessly promoting her new designs, eager to show people her work and realize her American Dream. She wanted to meet everyone . . . and everyone wanted to meet her. Who was this new Princess? How did royalty really live? Alas, all too many preferred to quiz her about her new family and the rich grandeur the privileged were supposed to sit back and enjoy. They wanted to hear about diamonds, not dresses!

It seemed as if no one was taking this young pregnant entrepreneur very seriously. Who was she kidding? Here was a Princess trying to sell jersey dresses stuffed into a suitcase! But she made the rounds anyway . . . albeit in a limousine.

No, Diane simply would not give up.

Two months after a much-adored son and heir was born to Diane and her Prince, the young mother decided it was time to get back to work. She wanted the magazines to support her. So, starting at the top, she made an appointment with the most important editor of all.

Red, red, red.

The office walls were dipped in the most ferocious shade of scarlet, a rather bold design choice that intimidated some but emboldened the brazen Princess. The room held its breath as she pulled samples from her bag to present to Fashion's Maharani. Two pencil-thin girls nervously slipped the dresses on, one by one, never once daring to take their eyes from their brooding monarch.

They were all waiting. Hoping. Praying for approval.

"Terrrrrific!" the Voice of Fashion suddenly boomed, her rolling Rrrr's fizzing like a pinball hitting the jackpot. "Terrrrrific! I say, how clever of you!"

Diane left dazed and bewildered. Her audience with the supreme arbiter of taste was over—and it had gone well. Very well. She sat crumpled on the floor of the outer corridor wondering: *What next?* How does one turn such *ringing* endorsement into *sales*?

Just then the answer came. Another fashion editor passed by and, having heard the reception Diane had just received, decided to help.

Looking at the samples she told her to book a suite and get ready for Fashion Week.

And with that Diane von Furstenberg's fashion line was born.

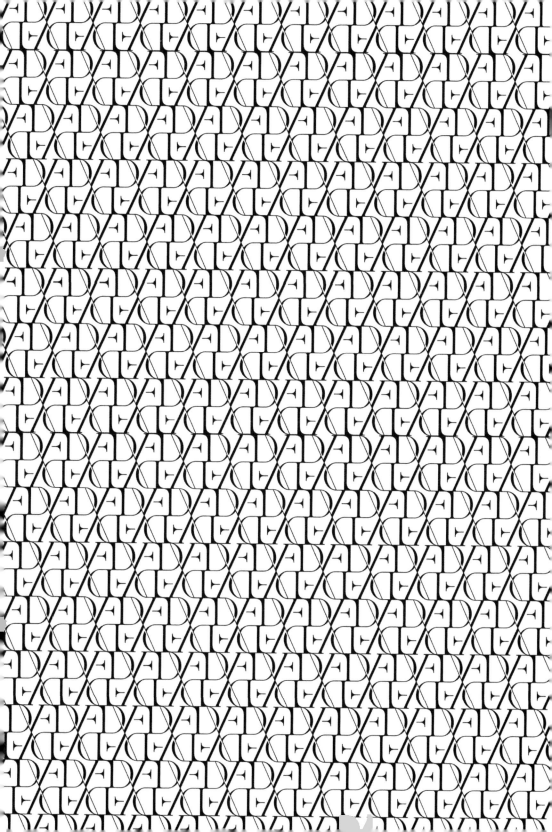

The young Empress felt close to experiencing a "fashion moment" of her own. She beckoned her new style gurus closer. "Come, gentlemen, let's see how we can work together," she smiled. The duo oozed and bowed their way toward her. They told the enthralled Empress how the cloth they made was far more special than any other, how the weave was more intricate and the fit more flattering than she could have ever dreamt . . .

Did they mention that this cloth also possessed a magical quality?

The Empress practically slid off her silken seat in rapt anticipation. Magical?! This was the answer to all her worries—would she look thinner? Taller? Bolder? Braver? It was too magically perfect to be true.

"This cloth is so special that only the wise, the beautiful, and the inherently chic can see it."

The Empress whooped with delight. Perfect—that would show her father that she had panache! Here was a fabric worthy of her mother that also came with the added bonus of weeding out all those who had no style. Excellent. She ordered enough fabric to create two dresses, a jacket, a coat, and a pantsuit. At last she felt satisfied she was covered for all possible moods and occasions.

What luck! This odd sartorial pair had come along just when she needed them most. She vowed never to judge on appearances again.

EUR

Diane was striking gold in her world, too. After a modest show, where models had walked the runway wearing her elegantly easy fashions and holding fresh tulips in their hands, orders for her dresses started to trickle in. This trickle quickly became a flood as the word spread and editors, models, and buyers all returned with friends. The seed of success had been sewn.

By night she was one half of society's golden couple; by day she was a mother, a wife and a hot new fashion designer. The art elite and fashion crowds were under her "spell" of sorts, and all wanted Diane to be seen and toasted with them at their events.

EKA!

Everyone was intrigued by the young entrepreneur Princess
who made these chic little go-everywhere dresses that
were so different from the elaborate, frankly cumbersome
ball gowns more associated with her chandelier-swinging
contemporaries.

It shouldn't have worked—but it did! Despite the fact that she
had no great master plan, and she still stored her dresses in
her dining room, there seemed to be no end to the trend in
sight. Everything about Diane was bold, exciting and new, and
this made her the name that *everyone* wanted to wear.

A year after the arrival of her son there was further joy when Diane gave birth to a beautiful bouncing baby girl. The young family's happiness seemed to be complete and this in turn was to inspire another creation.

THE BIRTH OF HER SIGNATURE DRESS . . .

The Prince was very proud of his young wife and offered his royal support to fund her designs, but Diane wouldn't hear of it. She was determined to do this without a royal check book and pawned a favorite diamond ring to cover factory costs before she began working on her next collection. Her dresses filled a space in women's wardrobes. Here were dresses for today that bridged the gap between the extreme fashions that the upper echelons of Paris, and high society, clamored to wear and the reality of most women's everyday lifestyle and budget. Diane had created the ultimate stylish solution.

Did dressing for the busy pace of everyday life mean you didn't care about looking good? Diane's answer was a resounding "No." She was creating garments that empowered women much like herself, giving them options that turned the workday uniform into something far more sassy and womanly.

These were clothes that *you* wore rather than clothes that wore you. Her motto was clear:

Feel like a woman, wear a dress.

And the message quickly caught on. The wrap dress leapt off the racks and into the wardrobes of every busy but style-savvy woman across America. Diane had created the very "look" modern women had been waiting for, and she toured the country to show them. She was causing a revolution! While other designers' clothes were expensive for the sake of being expensive, they often were not all that suitable to women's ever-changing lives.

Diane wasn't guessing or telling women what they wanted—she knew. She lived it, believed it, designed it, and this attitude made her designs right for every occasion.

Women weren't content to sit at home anymore; they wanted to be independent as well as feminine—to wear the crown as much as earn it. Diane was their heroine. She understood and embodied their ambitions and encouraged them to feel the same about themselves, too.

It takes a woman to know a woman.

Diane learned and listened to everyone she met—from the shop girl to the wealthiest guests at state dinners. She cared for them all with equal measure. And it showed in her fashions. Here was a woman who knew what a woman wanted and needed, who would make sure all this was stylishly wrapped together in a DVF dress. Diane von Furstenberg had truly arrived!

Back in the Imperial palace, the devious duo were also full of mirth. The Empress had fallen hook, line and sinker for their evil ploy. But, with the date of her mother's grand memorial drawing near, there was no time to waste—the young royal was anxious to get a glimpse of her new royal wardrobe.

"Guards!" she called, clapping her hands to command their immediate attention. Seconds later, the golden boys were tipped off the cushions at the foot of her throne and dumped unceremoniously onto the floor by the Empress's footmen.

"Come on, get to it," snapped the Empress. The duo snarled as they were ushered to the door. "We'll show this fool," they muttered to themselves. "Think she's better than us? Well just you wait and see!"

"Your Majjjjesty," the spindly one cooed, recovering his dignity to throw such a deep bow that his marabou-feathered hat trailed along the floor. "Before we leave we have but one request. . . . We'll need gold . . . I mean in order to spin your gown we will need golden thread . . . and plenty of it too."

The Empress, who had always refused to listen to anything as vulgar as talk of money, pursed her lips momentarily. She nodded to the Master of the Household, directing him to deliver the gold thread and anything else that glittered and might be of use. Her Imperialness was bored of this conversation—wasn't it obvious she didn't want anything cheap? Sadly, the Empress was so desperate to see her father smile again that she didn't realize what she had done. . . . She'd just handed over the crown jewels to common thieves.

At only twenty-six years of age, Diane was also at a crossroads. Business was booming, but alas, her fairy tale marriage had come to an end. Her union with the Prince had sadly become a casualty of her success.

She found herself at the start of a new chapter in life.

Diane was thankful for everything she had—her children, her fashion business, and now a budding new fragrance line—but she knew something was still missing. . . . She was living out her creative dreams, yet she was going home alone. Exciting as it was to make the cover of prestigious publications, most of all she wanted to find someone to share all her success with.

The first step to securing that balance was finding herself a haven where she had space to think and be herself. "Bluesky" became that perfect sanctuary, her very own private kingdom away from the fast pace and bright city lights.

With a new sense of calm Diane set out to embrace all she had created and view setbacks as challenges and keep her heart open and ready to love again. Soon after she did this Lady Luck smiled on her and a new suitor entered her life. This time he did not have the "classic" royal title, but he was HER Prince, her best friend, confidante and cheerleader.

Here was the Prince who held the key to the castle that was her heart.

This love story started off in the best way—the kind when you least expect it: they were friends. He was the youngest ever Chairman of a Picture House. Together the two adventurers shared their hopes and dreams for the future, and together Diane felt invincible.

To love and be loved is a very powerful thing. And though this was exactly what Diane had wished for, now was not the time this particular happy ending. . . . In addition to having beautiful children, a new love, and an ever-expanding business she had her soon-to-be-launched cosmetic line to think of. Only women know how to juggle everything, and only a great love is strong enough to set you free.

This love would simply have to wait.

Meanwhile the young Empress grew quite convinced that the new outfits she awaited would be the answer to all her prayers—finally the whole realm would see the successful woman she had grown up to be. Alas, she had pinned so much hope on the outcome, she didn't even notice how many high-ranking noses her new Fashion Favorites had put out of joint at court.

The devious duo had sensed that they were amassing enemies all around.

It came with the job—so they'd taken great pains to cover their tracks, smirking as they carried out their despicable plot.

Night and day, hour after the hour, they appeared to be working very hard. No one ever saw them stir from their studio—but the orders and deliveries flying through the door were surely a reflection of the activity within? If anyone *had* dared peek through the keyhole they would have seen just what kind of "work" it was . . .

Inside the pair were all too busy gorging on in-palace dining, surfing the Internet, and trying on the crown jewels: Hard Work. Their only concession to the "work" they did was to keep the volume high on their headsets so the music would drown out the constant patter of the accomplices—the poor tortured mice who ran endlessly up and down the looms in pursuit of cheese cruelly suspended just beyond their reach.

Despite the palace staff's best efforts to catch the duo lazing about, no one had yet succeeded. Anyone who passed along the corridor would hear the click-clack of the loom ceaselessly pulling back and forth, and then roll their eyes in despair.

The sound of the presumed toil grated on everyone's nerves. Indeed courtiers felt certain they'd soon be begging for mercy . . . if only the clickety-clack would stop. . . . Soon the entire palace was suffering from appalling migraines.

The Empress decided not to venture anywhere near the East Wing, where her protégés were installed. The official line was that "she didn't want to interrupt their creative process." The truth was, she didn't want a headache to ruin the weeklong string of luncheons and teas she had planned to host. She was so looking forward to telling her green-eyed groupies all about *her* latest fashion discovery.

Diane's life was spinning with activity, too. She was working and socializing more hours than seemed to fit in a day. This constant whirl often made her feel as if she too were running on an endless conveyor belt . . .

While the mix was fun and exciting, her dream of living a charmed life with just the right blend of work, play, love and family was quickly turning into the reverse.

Not one to give up easily, she decided it was time to press the "pause" button. Time to refocus exclusively on what would make her feel inspired, empowered and, most of all, happy.

Diane knew exactly where she could rediscover her *joie de vivre* — Paris!

Never to do things by half measures she embraced change inside and out. Diane chopped off her hair, found a marvelous boarding school for her daughter and created a new bohemian life for herself. It was liberating to leave the high life and madness of New York behind. She put away the spiked heels, traded her signature tux for classic tweeds and sensible sweaters and sought some much-needed peace. Here, in Paris, her soul could recharge surrounded by art, literature and love.

There was *no* peace to be had in the palace.

This was torture. The whole place was now sleeping with earplugs, and some even kept them in during the day.

CLICK CLACK, CLICK CLACK.

People of every rank were demented. Even the Empress was getting crotchety due to lack of sleep. This was serious. *When would her cloth be ready?* She summoned the Prime Minister and told him to check on the progress and report back.

Now, as you can imagine, most designers are very secretive about letting visitors view their process—but crooks are even more covert.

The Prime Minister tried to creep through the corridors, so as not to wake the sleep-deprived Kingdom, but the duo didn't care. They had set booby trap after booby trap along the route. As the PM tiptoed into the East Corridor, trumpets blared and fireworks popped all around him. He nearly jumped out of his skin. *When had all of this been authorized?* he wondered. But the answer didn't matter. The commotion was more than ample warning for the devious duo to hide the mice and merriment and get into position.

As the door opened, the designers barely looked up from their looms.

The Prime Minister peeked inside. But when he did he could scarcely believe his own eyes.

The duo had been working for weeks, yet he could see none of their progress. He had always thought of himself as a rather dashing gent—in his prime he had been considered by many to be a swashbuckling "Hollywood hero" type, and had nearly become a news anchor. But today he wasn't so sure of his sartorial savoir faire.

There appeared to be nothing on the loom.

I can't see anything, he worried to himself. *There's no fabric and yet I know I'm not stupid . . . This is not good . . .*

And with that he did the only thing he could do.

"Excellent work, gentleman," he commended them, clearing his throat. "Everything looks splendid. I shall trespass no longer. Absolutely beautiful. I can see the job is well in hand. I will report back to the Empress immediately."

The poor man snapped his smart polished heels together, turned, and ran.

The Prime Minister had barely pulled the door shut when the evil duo fell about laughing. They cackled so loud and so hard they had to stuff their caps over their faces to prevent anyone else from hearing. This was too easy. How silly and vain was this Kingdom? This was just the type of nonsense written in fairy tales. How could they possibly fall for this?!

Who cared?! They were onto a winner. The duo sent word to the Empress that a "major" celebrity had offered them double what she was paying to drop everything to go and work exclusively on her tour. They feared they might not have time to finish this royal commission . . .

The Empress barely lifted her eye mask. Triple the amount of gold, plus the keys to the kingdom, landed outside the studio threshold.

The new gold was a good incentive to continue their ruse, so they worked even louder, rather than harder, than before.

Soon all anyone could talk about was the Empress's New Look. Everyone was talking about this amazing fabric, and its mesmerizing powers. Even the two weavers started to believe the tale they had spun.

Priceless, thought the Empress triumphantly.

Just as speculation at court was reaching fever pitch, news of Diane's new venture in Paris was emerging too. Not only had she decided to write books, but Diane being Diane, she'd actually started her own publishing company, too.

She should have been very happy living this new lifestyle—with its long lunches, lazy Sundays and easy European pace. But there was just one catch. When she took off the rose-tinted glasses and looked around she wondered: *Is this really me?*

She felt very alone and could hear New York calling her to come home. She could also hear her mother's wise words echoing in her mind:

"We all do the same things; what makes us different is how we do them."

That's when she decided to do things *her* way again. It was time to confront life. To be true to herself.

Enough of this "time out." It was time to go home—she had a business, a name and a life to reclaim!

The young Empress had also made a decision: it was time for her long-awaited wardrobe to be ready. Her entire staff were utterly surly with tiredness, and she also was getting rather anxious about what accessories to style with her new look.

All the A-listers from near and far had been invited to the grand unveiling of her mother's statue. The Empress *had* to look her best. She wanted her father to notice her and to be proud of her. She also wanted to show the entire kingdom that she could equal the memory of this amazing woman.

"Send for the dressmakers."

The Empress snapped as she tried to down a slimming tea that tasted as bad as it looked. It resembled slime from the royal moat far more than an elixir of goodness.

The dressmakers were unaccustomed to being "summoned." They had grown rather arrogant during their time in the palace and had forgotten who was, in fact, Lord—or, in this case, Lady—of the Manor. Indeed the terrible two had become rather too comfortable treating the staff like they were their own, and to living as queens . . .

But that bubble was quickly burst. Reluctantly, they knew they must obey the command. The duo tugged on their livery and, with much grumbling, went to see the Empress.

"Well?" said the Empress, as the two deceivers bowed their way into her presence. This time there wasn't so much as a "hello." She simply rapped her fingers with impatience on the arm of her throne. Her good mood, and once welcoming royal nature, had all but evaporated.

"Empresses are not used to being kept waiting."

She was now shouting. "I have sent numerous Ministers and messengers, all of whom tell me you are 'working hard.' I've paid you a fortune in gold. I've had my staff cater to your every whim, yet I still have nothing to see?!" The Empress was furious and she towered over them like a head teacher scolding her two most naughty schoolboys.

To give the dressmakers their due, they promptly took off their caps, hung their heads in shame and bowed with the fury in her voice. Even though they were up to no good, the duo could see the Empress was not to be trifled with any longer.

"You told me I've been wearing out-of-date

68

rags!" she screeched, blushing at her own tone and at how much this mattered to her. "I asked you to help me and this is how you have repaid me? You treat me like a fool!"

The more spindly of the crooked pair gulped—surely this vain silly girl hadn't guessed? Their surveillance had been top-notch.

Luckily his partner in crime held his nerve and greasily stepped in. "*Majjjjjessssssty*, I don't know what the fools around you have been telling you, but while you have been getting thinner and even more lovely looking by the minute, we have been working through the night to weave something iconic enough to please our rrrroyalest muse," he simpered. "If you would just allow us to return to our studio we can finish the last bit of weaving today. After we have pieced the fabric together we would be honored to come back to attend to your personal fitting. . . . I know you will not be disappointed . . ."

"You have until sunset," the Empress replied curtly. "I will send my guards for you." Everyone in the room held their breath—the weavers knew that their time was up; no one had ever seen their ruler in such a rage.

The Empress was as furious as she was perplexed. Whenever she spent this sum of money on couture there was always *way more* sucking up, way more accommodating, and *way more* time spent on fittings.

She'd had enough of this "new" method of design.

"Tonight," she said, locking them both with her steely gaze. She was firmly back in the driving seat of this sale. "We will fit the outfit tonight. The celebration is tomorrow and I can't very well wear one of these other creations." Then she waved a dismissive hand towards the priceless rails of chiffons and laces that came from the A-Z of all the best designer names you can imagine that hung before her to prove her point.

"Everyone is talking about my 'New Look' and this incredible fabric. It's time you delivered." And with that she stormed out of the room, her guards, personal assistants, and terrified stylists all chasing behind her.

The day of the great celebration had finally dawned. But the Empress hadn't slept a wink. She had been feeling anxious ever since that fitting. Something felt very wrong but she just couldn't work out what . . .

The designers had marveled at how regal she looked. Her stylist had wept, "If only your mama could see you now." Everyone had assured her that she looked astounding—more beautiful than on any other occasion . . . yet she couldn't shake the feeling that she had been *naked* in their presence.

She had tossed and turned all night. Maybe she didn't have her mother's style? *Maybe she wasn't as chic or as clever as she had hoped to be?*

Why was she feeling so full of doubt?

The more she replayed the fitting in her mind, the more horrible it became. It also dawned on her that while she twirled in front of the mirror, none of her assembled courtiers had dared to make eye contact with her—or each other. Their behavior was most odd.

The Empress shuddered as she remembered standing before her own reflection. Dimpled, vulnerable, and all alone. No crown was going to hide that. The "fabric" seemed to show off *all* her insecurities—it was as if it was utterly invisible. What on earth did everyone else see?

She felt miserable. None of her attendants had so much as raised an eyebrow as she stood in all her finery. She cringed. The only comments she could muster after all those weeks of anticipation was "It's extremely light-weight, very comfortable."

Ugh. Maybe the fabric was right about her?

The dressmakers would be back in a few hours to help her get ready. (How could she do that herself if she couldn't even see the wretched outfit?!) She scrolled listlessly through her phone book. She needed her mum. And then she had an idea. She dialed a number she hadn't dialed in such a long time. She would call the Emperor. Her father was most certainly not an idiot.

Diane had butterflies about stepping back into the clutches of the fashion industry. But she had her family and so many friends thrilled to welcome her home. To make her triumphant return to New York complete she had reunited with her best friend and true soul mate, finding love once more with her Prince of the Picture House.

Why, she wondered, did she need to try her hand at fashion again? In the time she had been away she had changed and the business had, too. *Maybe I won't belong anymore?* she worried silently.

Then, just as she was about to cancel the showcase for her new collection, someone from her past caught her eye. It was one of her old friends and former buyers. Just when she needed it the most here was a sign that she should proceed.

"Diane!" the woman exclaimed, grabbing and hugging her like much-sought-after treasure. "Diane! Thank goodness! We need your dresses. All the designers are showing in the spirit of you—but why go for imitations when we can have the real thing? *We want you!*"

Diane paused. How many fairy wishes could one person be granted in life?

"I might just have something for you," she smiled, patting her shoulder bag. "Can you give me a few hours? Send your buyers round to my studio at this address."

Diane von Furstenberg was back in business!

The young Empress heard a light tap at the door.

"Darling, can I come in?" a distinguished voice asked on the other side.

"Daddy, is that you?" She could hardly believe he had answered her call, let alone traipsed all the way over to her private quarters. She hadn't seen him since . . . since . . . well, she couldn't remember when the two of them had last spent time alone together . . .

"OK, come in, but no peeking. . . . You can count to ten and then I'm going to show you my Super Duper New Look!

"Ten, Nine, Eight, Seven, Six, Five, Four, Three, Two . . . One and a half? One . . . Ready or not, here I come."

And with that, she stepped out from behind the screen, eyes shining, desperate to impress her father.

The Emperor blinked once, then blinked again before roaring with laughter. "Oh! You are a tonic!" he chuckled. "How I've missed you! Keep the fan there!" His crown rolled off his head and clattered to the floor as he shook with an unrelenting case of the giggles, his ruddy cheeks growing even redder than usual. "Goodness, I've been so wrapped up in my own grief I forgot how to laugh—or rather how much you make me laugh, even more than your mother ever did, minx!"

The Empress laughed too as she watched tears of mirth fall down her father's face. "OK, but go and put some clothes on now," he said to her—after all that's why he was here.

"WHY ARE YOU COMPLETELY NAKED?!

"Is this a new trend? The latest fashion statement? If so, I'm not sure it would work for me and the royal belly . . ." Then he was off again, guffawing with laughter so loudly that one of the Empress's attendants cautiously popped her head round the door to see what all the commotion was about.

The maid wouldn't have believed it had she not seen it for herself, but there, before her very eyes, the Imperial family lay cracking up until their sides hurt . . .

Until, all of a sudden, the Empress fell silent, realizing the truth of the situation for the first time. She was stark naked.

"Daddy, are you sure you don't see anything?

"These are my new clothes! Only those that are wise can see them . . ."

she stammered, knowing in her heart as she spoke that this was not so. But she boldly continued, while covering her modesty with crown, fan and some quickly snatched necklaces. "I know I will never equal Mama in the fashion stakes, but these new designers came to me and assured me that I was their muse and that only the very, very clever would be able to see the outfit they had created for me. . . . Do you really see nothing? I've paid a whole year's allowance of gold for it."

At this the Emperor stopped laughing and raised his brows. "What?" he roared.

"Oh don't be cross with me, Daddy, please don't be cross with me," the Empress stammered. "I know I will never be her, I will never inspire artists to sculpt statues like she did, but I thought, well . . . I hoped that if I at least stood out you would notice me again . . . maybe even want to talk to me . . . I just wanted you to see all I had done, all the work I had achieved . . ." Her voice wavered and tears started to trickle down her cheeks.

Never had the Emperor felt more ashamed. He took off his regal mantle and draped it around not an Empress but his sobbing child. How had he let this happen? Hadn't he always vowed to be a father first, an Emperor second? He had let his daughter and his wife down when they needed him most.

She was their greatest achievement.

His heart felt like it might break as his most precious jewel dissolved into tears and clung to him, finally able to relax as her daddy was back to look after her.

"Daughter, can you ever forgive an old fool for my selfishness? You don't need labels or a crown to prove what an amazing woman you have become. You are perfect—you always have been. I love you just the way you are . . ."

"But what about Mother's statue?" The Empress wailed. "I've got nothing to wear to see Mummy."

"No, looking like this, you certainly don't. But luckily we've had this private little 'undressed' rehearsal. Can you trust your old dad—it just so happens that I have the perfect outfit for you . . ."

"Manuel!" he called, as his harried footman fell through the curtain. "Did you hear all that? I think I would like to meet these 'marvelous' dressmakers for myself. Tell them I want to see them. Throne Room. Pronto. Call the executioner too. . . . Oh, and Manuel . . . don't let on that I know." Then he gave his daughter a stage wink. "Tell them the Emperor would also like a suit, in the same fabric. We want to be wisely matched at the ceremony today . . ."

The footman picked up the phone, frantically ticking off the list of things to be done. The guards, the thrones, the red carpet, the executioner were all on standby. Now to summon those designers, he would do that in person. He bowed and made his exit as the Empress gladly wiped the last of her tears away. A sense of calm finally swept over her. It was nice not to have to make all the decisions for once. She threw on some of her old clothes—cashmere had never felt so good—and then flung open all seven doors of her wardrobe in a panic, shrilly exclaiming that she *still* had nothing to wear, much to her father's amusement.

Back in the city that never sleeps . . .

Diane was running through plans for her upcoming show at fashion week when her phone rang. There were the invites, the seating plan, the model castings and the fittings as well as the pre-show interviews and appointments to arrange.

It was around this same time that the dastardly designers were summoned to see the Emperor.

Diane had seen what was being touted as haute couture these days, and was as mystified as anyone by its appeal. Why would women want to wear some of these clothes? Did they really empower and make a new generation feel sexy? She somehow doubted it.

The phone call was a welcome interruption from her musings, and she was especially glad to discover it was a friendly voice from the past on the other end of the receiver. Once she heard about the Emperor's dilemma she quickly grabbed an armful of new dresses, stuffed them in her faithful shoulder bag, and was off, happily swapping the madness of yet another fashion week for what sounded like the utter chaos of court . . .

The tricky twosome were equally delighted with their latest Imperial summons. Think of all that gold! What a coup! Fancy conning both the Empress and her father. They would be able to retire wealthy and infamous throughout the land if they pulled this off. Everyone had heard about this incredible new outfit, and everyone was curious to see if they were wise enough to appreciate its magic powers. What an evil triumph this could be!

The Emperor and his daughter sat united, side by side, as the wicked pair entered the room. "Your Majesties," they simpered as they approached, sweeping the chamber floors with their grandest bows.

"Seize them at once," the Emperor commanded his guards barely glancing in their direction.

"OFF WITH THEIR HEADS."

He snapped the neck of the gingerbread man he had been eating as if to demonstrate.

"Majesty?" cried the terrified designers, but the Emperor didn't reply. His mouth was full of delicious gingerbread morsels, and besides, he was already bored of these scoundrels. He waved his wrist for them to be taken away with hardly a care about what fate befell them after the way they had tricked his daughter. He dusted the crumbs from his robe and instead reached for another cookie.

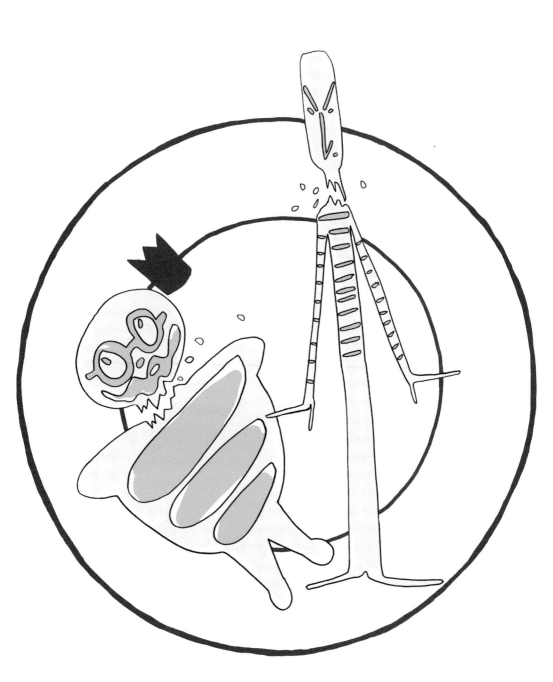

"Father, I'm so ashamed. My vanity was the cause of all this . . ." the Empress began.

"There, there my child," her father whispered softly. But as he soothed her, his attentions were diverted to the striking woman who had just strode into the room wearing the spikiest of high heels.

"Diane!" he beamed as he rose to hug her in a far warmer manner than seemed Imperial. "My dear friend, I am so glad that I ran into you at your son's wedding. This is my darling daughter. Do you think you could help us? Bit of a 'fashion crisis' here, I think you would call it."

Diane held her hand out toward the Empress and shooed all the men from the room—including the Emperor. Rather than be offended by her take-charge attitude, he was really rather amused. "*Women,*" he smiled to himself; he had missed them. Diane watched as the young girl dried her eyes. She looked so like her mother—Diane's dear childhood friend with the long blond hair. "Your mother used to live in my designs," she explained. "Try this, it's just a simple wrap, but it has some magic all of its own . . ."

Then Diane applied the child's makeup and, before long, it was time for the trio to set off to see the statue. Today they would celebrate and remember all that was wonderful about their dear friend, wife, mother and Empress.

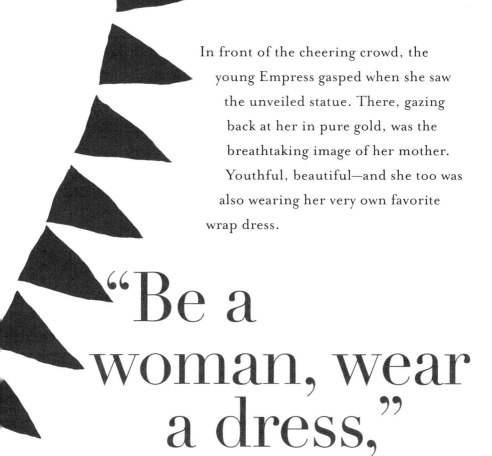

In front of the cheering crowd, the young Empress gasped when she saw the unveiled statue. There, gazing back at her in pure gold, was the breathtaking image of her mother. Youthful, beautiful—and she too was also wearing her very own favorite wrap dress.

"Be a woman, wear a dress,"

the statue seemed to whisper as the daughter raised her head to softly kiss its golden check.

Once again Diane's wrap had saved the day, and was now being discovered by a whole new generation with the same needs, worries, desires, and dreams as the last.

DIANE von FURSTENBERG

TIMELINE

1970s

In 1970, Diane von Furstenberg arrives in America, a newlywed princess with a trunk full of printed jersey dresses she had designed and made at a textile factory in Italy. After receiving the ultimate blessing from *Vogue* editor-in-chief Diana Vreeland, in 1972 Diane shows her first collection at New York Fashion Week. In 1974 she combines her already signature top and skirt with iconic results, and the wrap dress is born. By 1976 she has sold more than five million of her signature dresses and landed on the cover of the *Wall Street Journal* and *Newsweek*, the latter calling her "the most marketable woman since Coco Chanel." Diane von Furstenberg has become a brand!

1980s

In the 1980s, after the successful launch of her bestselling fragrance, Tatiana, Diane opens her first luxury boutique at the Sherry-Netherland Hotel, on New York's Fifth Avenue, to house her couture collection. In 1985 Diane sells her company, takes a hiatus from fashion and moves to Paris, where she establishes a French publishing house called Salvy. But by the decade's end, New York and the fashion world are calling her home.

1990s

In 1997, inspired by a new generation of hip young girls who are buying the original wrap dresses in vintage shops, Diane relaunches her collection. She celebrates her fresh start in the industry by writing a business memoir, *Diane: A Signature Life,* and establishes her headquarters in New York's West Village.

2000s

Now Diane has secured her status as the ultimate "Comeback Kid," and the 2000s are all about expansion. The brand expands to a full ready-to-wear collection; DVF expands its retail presence around the world and onto DVF.com. New collections are launched: swim and beach, eyewear, and a line of fine jewelry for H. Stern. Philanthropy efforts expand to include Vital Voices, International Women's Day, and The DVF Awards. DVF once again epitomizes effortless glamour and confidence.

2010s

Diane continues to take the DVF brand to new heights around the globe, connecting her vision shown on the runway at New York Fashion Week, embracing new collaborations, and leading the way with new fashion technology communications.

Diane commemorates her four decades in the business by creating the "Journey of a Dress" exhibition, which shows in Moscow, São Paulo and Beijing. The retrospective spans her career and celebrates Diane's reach as a designer, art muse and collector, featuring over 80 vintage and contemporary dresses, original artwork and personal mementos.

Still renowned for its iconic wrap dress and signature prints, DVF has carried this practical approach to glamour and gone on to create an extensive array of cult accessories and handbags, from tech-savvy cases to party-ready clutches. In 2012, DVF has evolved into a global contemporary luxury lifestyle brand sold in over 70 countries with over 60 freestanding shops worldwide.

For more on Diane's world, log on to www.DVF.com.

Feel like a
woman,
Wear a dress!

Diane Von Furstenberg

OFFICE OF THE
Editor-in-Chief

VOGUE

689-5900

THE CONDÉ NAST PUBLICATIONS INC.
420 LEXINGTON AVENUE, NEW YORK, N. Y. 10017

April 9, 1970

Diane:

I think your clothes are absolutely smashing.

I think the fabrics, the prints, the cut are all great. This is what we need. We hope to do something very nice for you.

Also, do you need any help with stores?

If there is anything you want us to do, please don't hesitate to call.

Very sincerely yours,

much love!
Congratulations —

Diana Vreeland

Diana

Princess Egon Furstenberg
1155 Park Avenue
New York,New York

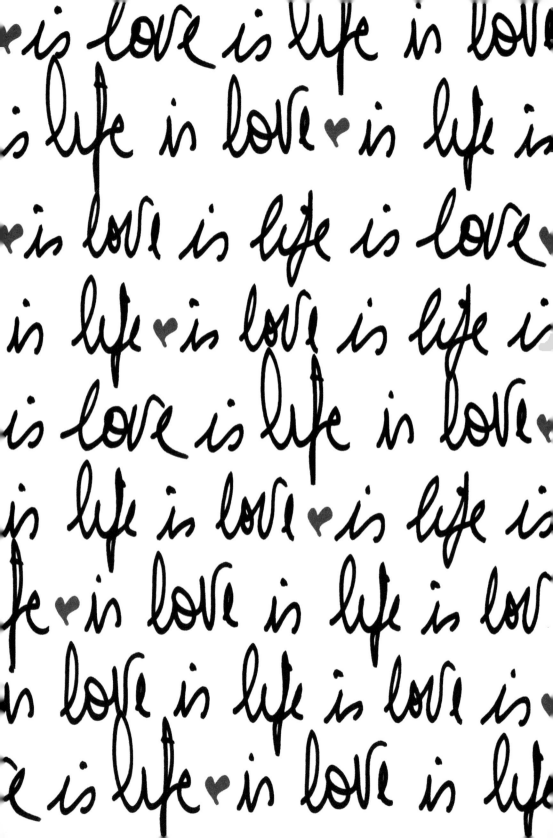

ACKNOWLEDGMENTS

To Diane, thank you for embracing this fairy tale with such generous enthusiasm and passion. I am truly grateful for all your grace, guidance and creative eye that transformed the Empress in this book, and all of us.

Thank you to everyone at Diane von Furstenberg, most especially to Neil, Grace, Daniel, Kristen, and all the DVF team who believed in this project.

Thank you to my publishers at It Books, to Carrie, Hope, Lorie, Kevin, Mary Beth, and Signe and the amazing team there who are so dedicated to creating, promoting, producing and publishing this series—especially Cal Morgan, Joseph Papa, and Andrea Rosen for their magic.

Most of all this book is for all the women out there who are fabulous, with or without a crown; this is for you with the hope it will inspire you to be fearless and make those dreams come true, just like Diane would.

May Diane inspire you to find beauty in your own life, both inside and out. Whether you find beauty in fashion, love, or fairy tales, may your spirit always be generous and may you always give back.

*it***books**

HarperCollins books may be purchased for educational, business, or sales
promotional use. For information please write: Special Markets Department,
HarperCollins Publishers, 10 East 53rd Street, New York, NY 10022.

FIRST EDITION

Designed by Lorie Pagnozzi with Camilla Morton

Library of Congress Cataloging-in-Publication Data is available upon request.

ISBN 978-0-06-191732-5

12 13 14 15 16   RRD   10 9 8 7 6 5 4 3 2 1

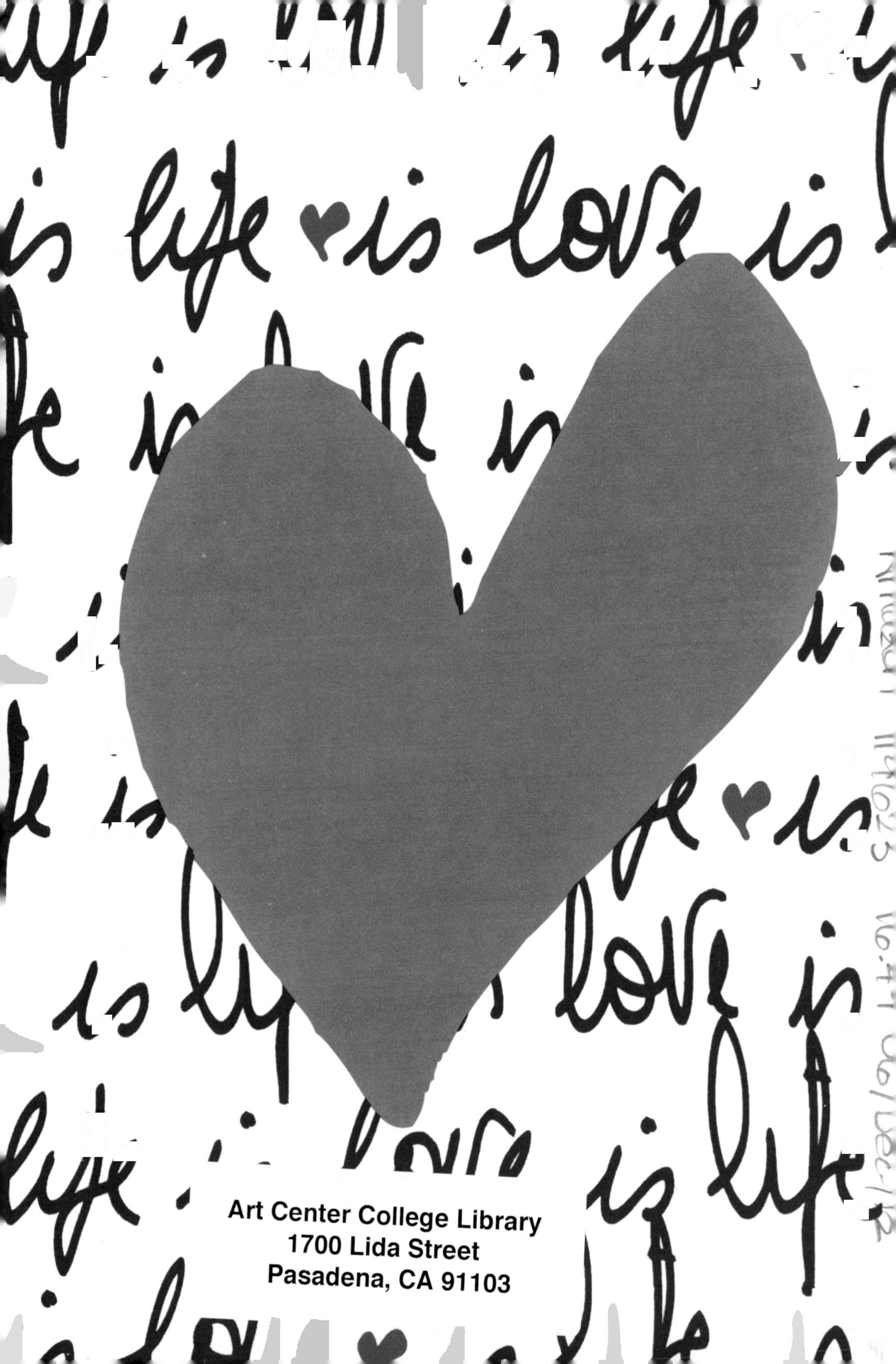
is life is love is